YOUR AMAZING BODY

YOUR AMAZING HEART

DWAYNE HICKS

New York

Published in 2023 by The Rosen Publishing Group, Inc.
29 East 21st Street, New York, NY 10010

First Edition

Editor: Greg Roza
Designer: Michael Flynn

Photo Credits: Cover, p.1 Yakobchuk Viacheslav/Shutterstock.com; p. 5 Prostock-studio/Shutterstock.com; p. 7 SciPro/Shutterstock.com; p. 9 mybox/Shutterstock.com; p. 11 Lightspring/Shutterstock.com; p. 13 Designua/Shutterstock.com; p. 15 Jihan Nafiaa Zahri/Shutterstock.com; p. 17 Alila Medical Media/Shutterstock.com; p. 19 iVazoUSky/Shutterstock.com; p. 21 fotoliza/Shutterstock.com.

Cataloging-in-Publication Data

Names: Hicks, Dwayne.
Title: Your amazing heart / Dwayne Hicks.
Description: New York : Powerkids Press, 2023. | Series: Your amazing body | Includes glossary and index.
Identifiers: ISBN 9781725339613 (pbk.) | ISBN 9781725339637 (library bound) | ISBN 9781725339620 (6pack) | ISBN 9781725339644 (ebook)
Subjects: LCSH: Heart–Juvenile literature. | Cardiovascular system–Juvenile literature.
Classification: LCC QP111.6 H54 2023 | DDC 612.1'7–dc23

Manufactured in the United States of America

CPSIA Compliance Information: Batch #CSPK23. For Further Information contact Rosen Publishing, New York, New York at 1-800-237-9932.

CONTENTS

Let's Get Pumping!

Did you know your heart is a **muscle**? It's a muscle that works very hard. In fact, it never stops working! Your heart is a **pump** that moves blood throughout your body. Blood carries things you need to live.

Where Is Your Heart?

Your heart is about the size of your fist. It's located to the left of the center of your chest. Your heart is made up of four parts called chambers. The chambers work together to send blood around your body.

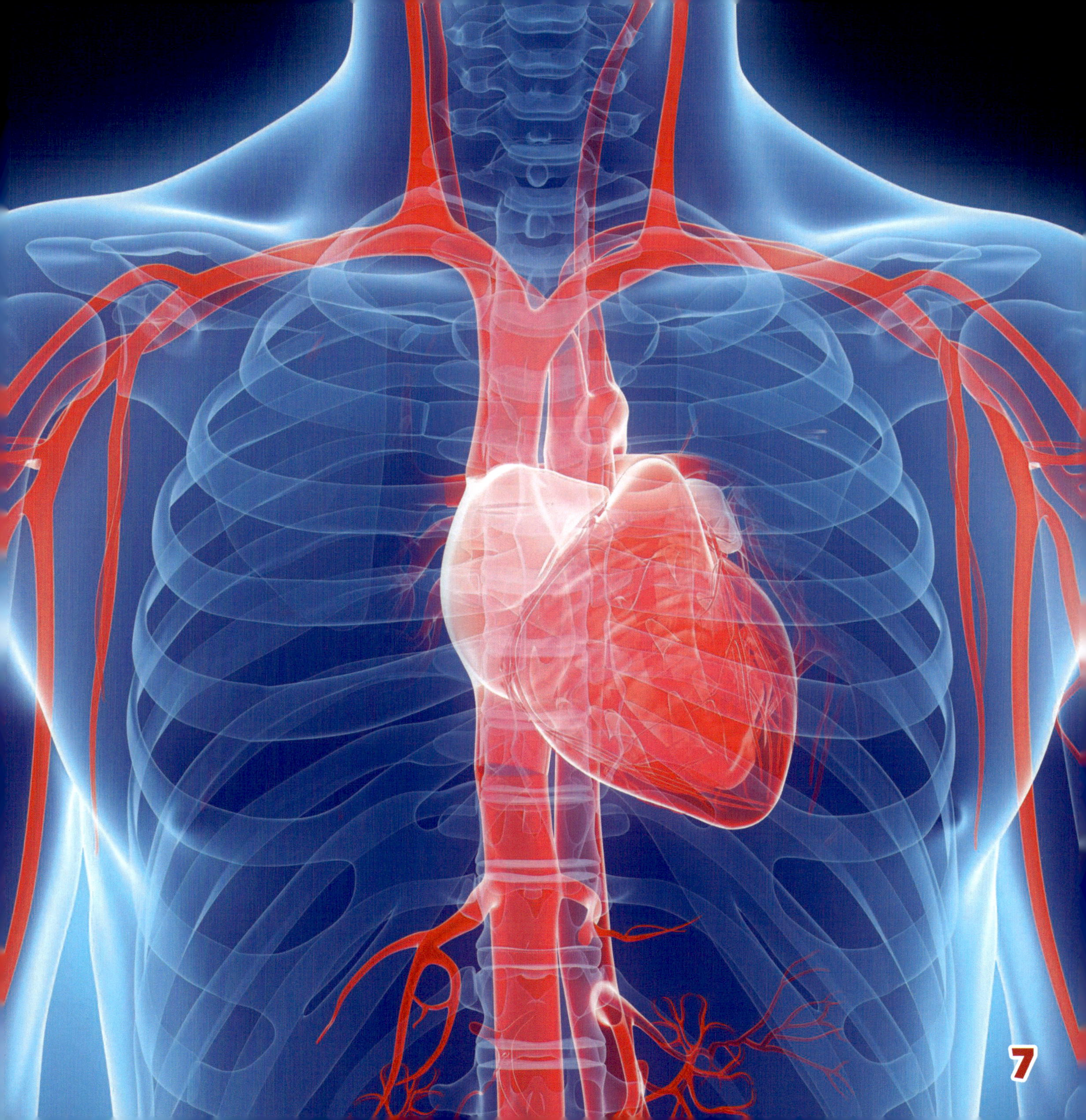

The Circulatory System

Your heart is part of your circulatory **system**. This system also includes blood vessels. Blood vessels carry blood throughout your body. There are two kinds of blood vessels. Arteries carry blood away from the heart. Veins carry blood back to the heart.

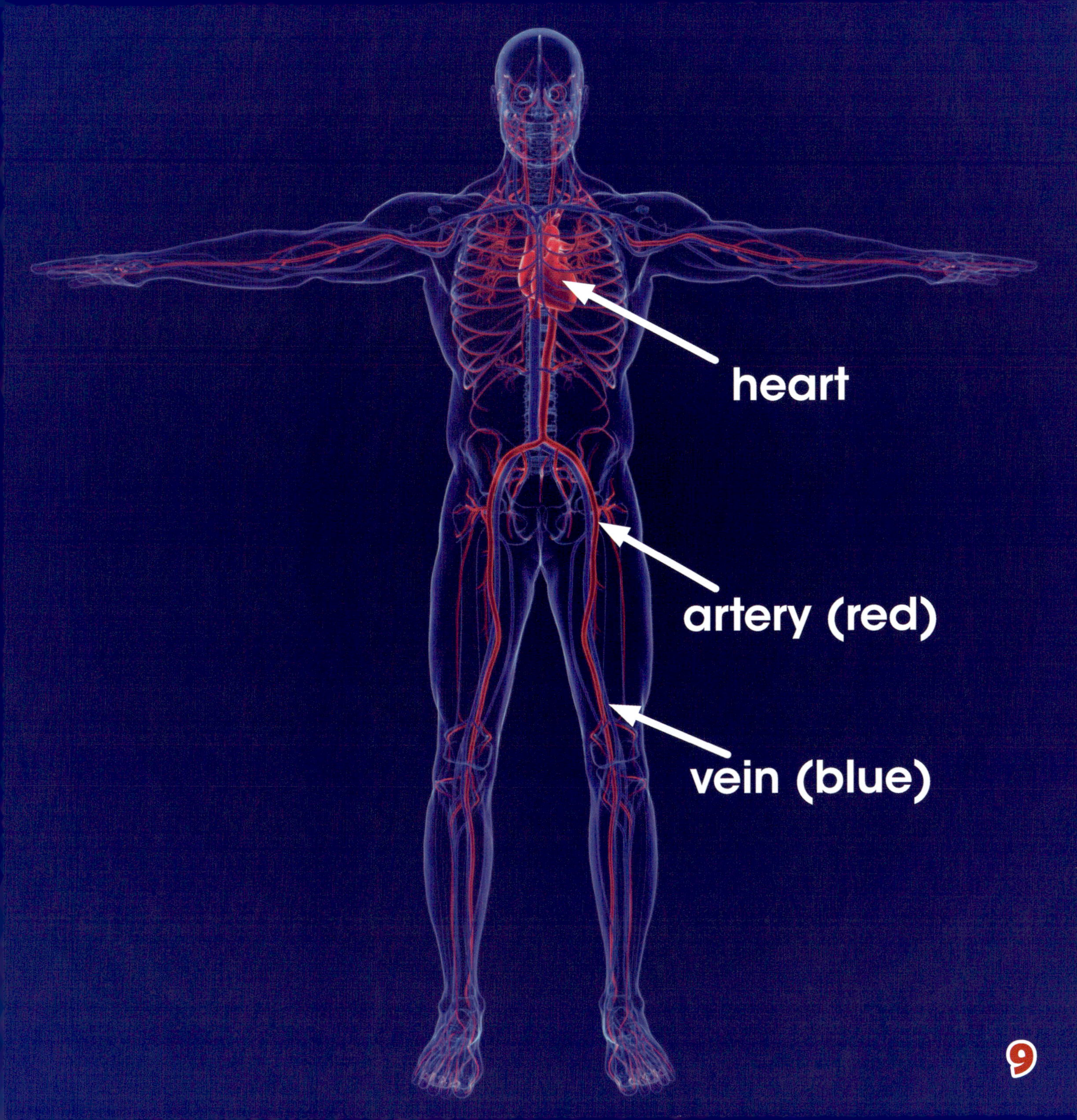
heart
artery (red)
vein (blue)

Moving Blood

The heart pumps blood through the body and makes sure body parts get the **nutrients** they need. Blood carries the gas **oxygen** from the **lungs** to body parts. Blood also carries the gas **carbon dioxide** away from body parts and to the lungs.

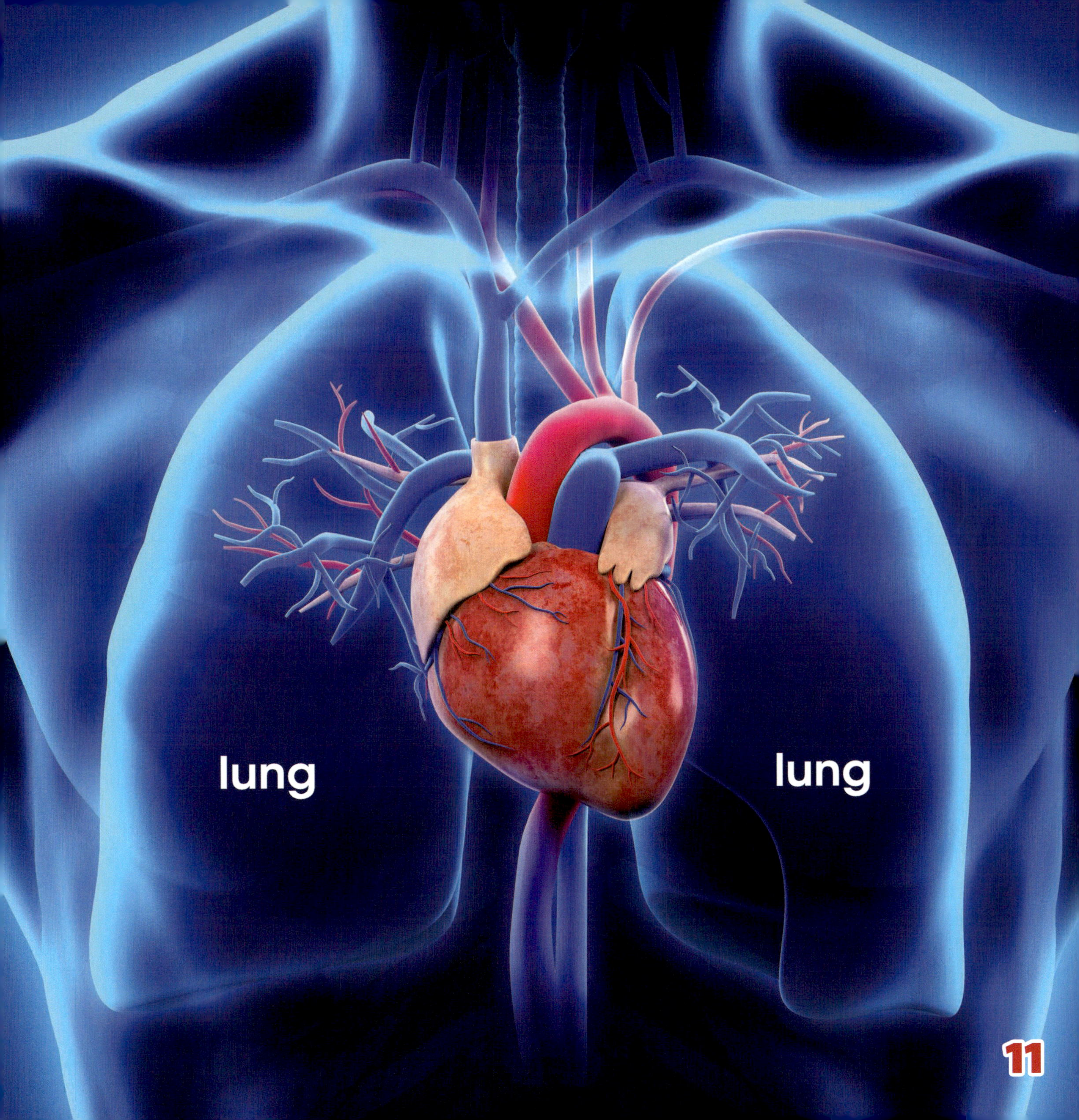
lung
lung

How the Lungs Help

The right side of your heart takes blood with oxygen from the lungs. It then sends the blood through your arteries. The left side of your heart takes blood with carbon dioxide from the body. It then sends the blood to the lungs.

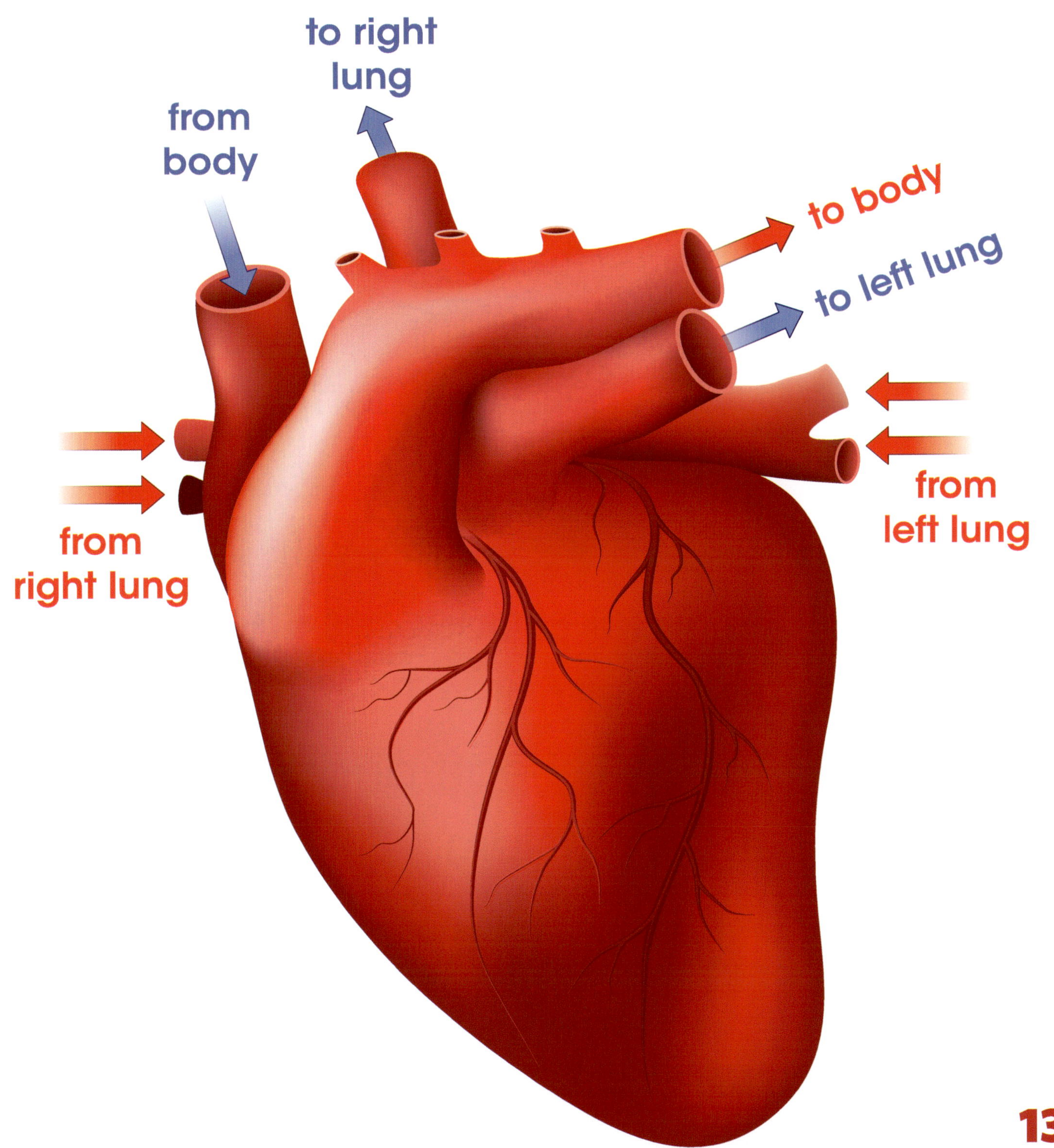
to right lung
from body
to body
to left lung
from right lung
from left lung

Heartbeats

Have you ever felt your heart beat in your chest? Just before each heartbeat, your heart fills with blood. Then, the muscle squeezes. The chambers get smaller. This forces blood out of the heart and through the circulatory system.

Open and Close

You might wonder why blood only flows one way through the circulatory system. Your heart has four special parts called valves. A valve is like a door that opens and closes. Valves open to allow blood through. They close to stop blood from flowing backwards.

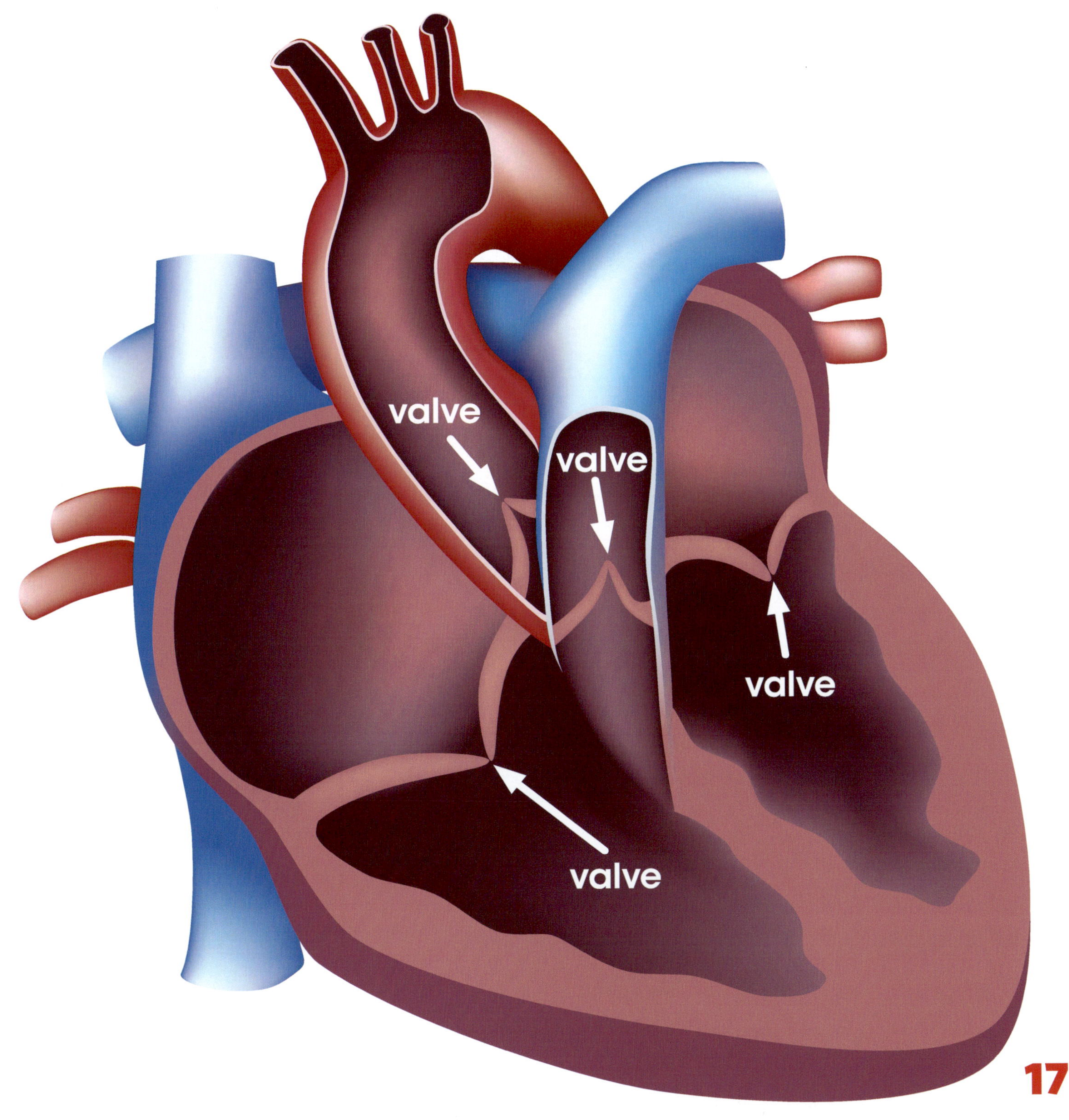
valve
valve
valve
valve

Heart Illnesses

There are illnesses that can affect the heart and circulatory system. Some people are born with heart illnesses. Other illnesses can occur because of the way a person lives. This could mean they don't get enough exercise or don't eat healthy foods.

Heart Health

An adult's heart beats between 60 and 100 times a minute. That's between 3,600 and 6,000 times an hour! To keep your heart healthy, get exercise every day. Make sure you mostly eat foods that are good for the heart and blood vessels.

GLOSSARY

carbon dioxide: A gas our bodies create as waste and that is removed from the body through the lungs.

lung: One of two body parts that exchanges carbon dioxide for oxygen when we breathe.

muscle: A part of the body that produces motion.

nutrient: Something taken in by a plant or animal that helps it grow and stay healthy.

oxygen: A gas our bodies need to live and grow that is taken in through the lungs.

pump: Something that moves liquids or gases from one place to another. Also, the process of moving a gas or liquid.

system: A group of body parts that work together to perform an important function in the body.

FOR FURTHER INFORMATION

BOOKS

Edwards, Jonas. *The Circulatory System*. New York, NY: Gareth Stevens, 2021.

Rose, Simon. *Circulatory System.* Calgary, Alberta, CA: Weigl, 2019.

WEBSITES

Cardiovascular System
www.ducksters.com/science/blood_and_the_heart.php
Find out more about the circulatory system (also known as the cardiovascular system) at this detailed website.

How Does the Heart Work?
mocomi.com/how-does-the-heart-work/
Read much more about the heart and see an animated video explaining how it works.

Publisher's note to parents and teachers: Our editors have reviewed the websites listed here to make sure they're suitable for students. However, websites may change frequently. Please note that students should always be supervised when they access the internet.

INDEX